PRESCHOOL MATH AT HOME

PRESCHOOL MATH AT HOME

SIMPLE ACTIVITIES TO BUILD THE BEST
POSSIBLE FOUNDATION FOR YOUR CHILD

KATE SNOW

WELL-TRAINED MIND PRESS

Reprinted February 2025 by Versa Press
7 8 9 10 11 12 Versa 30 29 28 27 26 25 Job #J25-01078
Publisher's Cataloging-In-Publication Data
(Prepared by The Donohue Group, Inc.)

Names: Snow, Kate.
Title: Preschool math at home : simple activities to build the best possible
    foundation for your child / Kate Snow.
Description: Charles City, VA : Well-Trained Mind Press, [2016]
Identifiers: LCCN 2016930340 | ISBN 978-1-933339-91-7
Subjects: LCSH: Counting--Study and teaching (Preschool) | Numbers, Natural--
    Study and teaching (Preschool) | Arithmetic--Study and teaching (Preschool) |
    Educational games.
Classification: LCC QA135.6 .S66 2016 | DDC 372.7--dc23

Cover design by John Hamman
Diagrams by Debra Pearson
Illustrations by Jeff West

TABLE OF CONTENTS

PREFACE

Five years ago, when my first child was a preschooler, I marveled at how quickly and easily he learned new things. Whether he was building block towers, looking at his favorite picture books, or digging in the sand box, he always seemed to be absorbing new information like a sponge. His vocabulary grew every day as he learned to name and describe everything around him.

I wanted my son to learn about numbers and the language of math with as much interest and excitement as he learned about the rest of the world. But when I looked for a preschool math program, none of the available options captured the joy and enthusiasm for numbers that I was hoping to instill. Most programs were workbooks with lots of repetitive matching activities and far too much writing. I knew my son would learn math best if he could move, talk, and play—not sit in a chair and do worksheets! I didn't want him to think of math as boring pencil-and-paper work, but as a natural part of everyday life that he could use to better understand the world around him.

Since I couldn't find a high-quality, developmentally-appropriate preschool math curriculum, I decided to create my own. I had plenty of background as an elementary math educator—I had majored in math and teaching in college, taught fifth grade, and even written math curricula—so I read everything that I could find about how preschoolers learn math and created simple, purposeful activities that would give my son a thorough understanding of the numbers from zero to ten. I was busy with a new baby, though, so every activity had to take less than ten minutes, be easy to implement for my sleep-deprived self, and use things I already had around the house!

My son and I had a wonderful time playing with numbers together while his baby sister napped. Now, five years later, I can see how the activities we did together developed his confidence and gave him the skills he needed to thrive in math. Math is now his favorite subject, and he loves to tackle challenging problems. His little sister is now five, and she's learned preschool math using the same activities that I used with her brother. As she begins kindergarten math, I'm seeing how the foundation laid by our math time together has helped her become a confident and enthusiastic math student as well.

I'm thrilled to be able to share with you this preschool math curriculum, based on the simple, straightforward activities that I used with my own kids. Even if you don't feel very comfortable teaching math, this book is designed to give you the tools you need to give your child a great start in math. Every activity is clearly explained, step by step. Notes throughout the book help you understand the reasoning behind the activities: what skill each activity develops, why they are sequenced in this particular order, and how each activity will help your child develop solid math skills. By the time you reach the end of the book, your child will be ready to learn kindergarten math with confidence.

I wish you and your child much fun and joy as you explore the world of numbers together. Happy Math!

Kate Snow
Grand Rapids, Michigan

INTRODUCTION

What Your Child Will Learn

Young children develop many informal ideas about numbers long before receiving any explicit instruction. For example, most one-year-olds can tell the difference between one cookie and three cookies (even if the only word they have to express the difference is "Mine!").

The activities in this book will build on what your child already knows and help her become skillful with the numbers from zero to ten. They will also teach her the language of math, the words and written numerals that allow us to communicate about numbers and quantities. As you use the activities to teach your child, you'll focus on five key skills that help preschoolers develop a solid math foundation:

- Counting
- Subitizing (recognizing quantities without counting)
- Recognizing written numerals
- Comparing numbers
- Joining and taking away (beginning addition and subtraction)

Let's take a closer look at each of these skills.

Counting: More complicated than adults realize

Counting is the foundation of preschool math. As children count, they learn what numbers *mean*: that two can refer to two apples, two books, two sisters, two jumps, or two taps. The idea that numbers represent quantities is very abstract; counting is what makes this idea concrete for young children.

Counting seems basic, but young children's counting mistakes show what a complex skill it is. Perhaps you've seen a small child very earnestly point to a pile of crayons and say "one, four, three, eight!" Or, you may have watched a child "count" by pointing in the general direction of a pile of blocks and rattling off "onetwothreefourfivesixseveneight" without any attention to how many blocks there actually are. Some children can even point to three toy cars and say "one, two, three"—but when asked, they can't say how many toy cars there are!

To be able to count accurately, a child has to learn:

- The order of the counting words ("one, two, three," etc.),
- That you have to count each item once and only once,
- That you can count objects in any order,
- That any kind of thing can be counted (even intangible ones like sounds or jumps),
- That the last number said when counting is the total number.

Since preschoolers are concrete thinkers and learn by doing, you can't just *tell* them these principles, though. To understand these important ideas, they need plenty of experience counting real objects, pictures, jumps, and sounds. The counting activities in this book are designed to help your child develop this thorough understanding of counting.

Your child will learn to count to five in Chapter 1. In Chapter 2, she'll extend the counting sequence further and learn to count to ten.

Subitizing: Recognizing quantities and combinations without counting

While counting is the foundation of preschool math, it's essential that children also learn to recognize groups of items *without* counting. This is called learning to *subitize* (pronounced SOO-bi-tize). To understand better what subitizing is, take a quick glance at these illustrations.

Notice how you can immediately tell that there are four fingers raised, without counting each finger? Or that you can just "see" that there are three stars? That's subitizing. Most adults can easily recognize up to five items at a time, no matter how they are arranged.

Once there are more than five items, though, it's much more difficult to determine the quantity without counting—unless the items are organized into smaller groups. Take

Can you tell at a glance how many fingers and stars there are?

a quick glance at each of the illustrations below. Can you immediately tell how many objects are in each box, or do you have to count them?

Each box contains eight objects. You can probably tell at a glance that there are eight triangles and eight squares, because those objects are organized into smaller groups (four and four triangles; five and three squares). But since the dots are scattered randomly, it's very difficult to tell how many there are without counting each dot one-by-one.

In Chapter 3, your child will learn to recognize quantities from zero to five by sight. You'll use a simple grid of five squares (called a five-frame), along with your fingers and household objects, to teach her this skill.

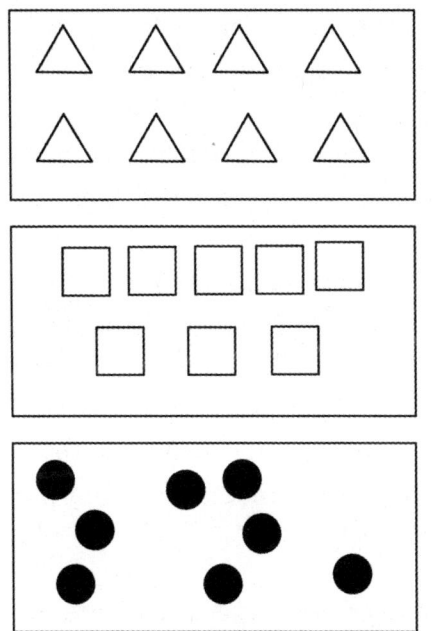

Can you tell at a glance how many triangles, squares, and dots there are?

Once she has learned to recognize small quantities by sight, she'll then use this skill to learn the combinations that make five (four and one; three and two; five and zero). The five-frame makes it easy to

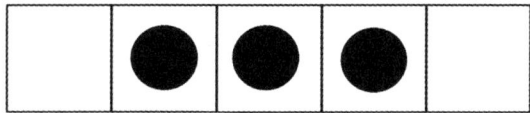

Five-frame with three counters

learn these combinations visually; in the example to the right, three boxes are full and two are empty, so three and two must make five. Learning these combinations previews addition and subtraction: a child who can mentally "see" that two and three make five will later be able to solve simple addition and subtraction problems like 2 + 3 or 5 − 2 with ease.

In Chapter 4, you'll use a ten-frame (a grid of *ten* squares) to teach more complex subitizing. Your child will learn how to recognize six to ten objects as combinations of five and some more. For example, this arrangement on the ten-frame will lead your child to discover that seven equals five and two more (since there are five circles on the left-hand side of the dark line, and two circles on the right-hand side).

Just as the five-frame helps children learn the combinations that make five, the ten-frame will help your child learn the combinations that make ten

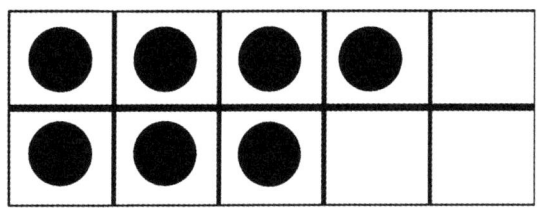

Ten-frame with seven counters

(five and five; six and four; seven and three; eight and two; nine and one; ten and zero). With the ability to visualize combinations, she'll be well-prepared to master addition facts in kindergarten and first grade.

Recognizing written numerals: Making the connection between concrete objects, spoken words, and written symbols

Written numerals (0, 1, 2, 3, etc.) are our short-hand for communicating about numbers. After children have a firm understanding of the concept of numbers and have gotten to know the numbers from zero to ten well, they are ready to connect spoken numbers with written symbols. Learning the written numerals in preschool will make calculations and more advanced work in kindergarten much easier. (Imagine trying to solve 3 + 6 when you're not quite sure what the symbols "3" or "6" mean!)

Recognizing these written numerals is much like learning the letters of the alphabet, because children learn to match a symbol with a spoken word. Just as learning the alphabet takes time, your preschooler will likely need quite a bit of practice to learn which squiggly shape goes with which spoken number. Chapter 5 focuses on teaching your child how to recognize the written numerals from 0 to 10.

Comparing numbers and quantities: Which has more?

Learning to compare quantities helps preschoolers begin to make sense of the relationships between numbers: seven is one *less* than eight, but it is one *more* than six. Your child will build on these relationships in kindergarten addition and subtraction. For example, a kindergartner might use her knowledge that eight is one more than seven to solve 7 + 1.

Preschoolers already understand the concept of more and less informally, especially if they feel that someone else is getting "more" and they are getting "less"! Even without any instruction, most are able to compare quantities if only small amounts are involved, or if two quantities are very different from each other visually. For example, in the pictures below, your child could probably tell right away which box of cars has more cars, or which plate of cookies has fewer cookies.

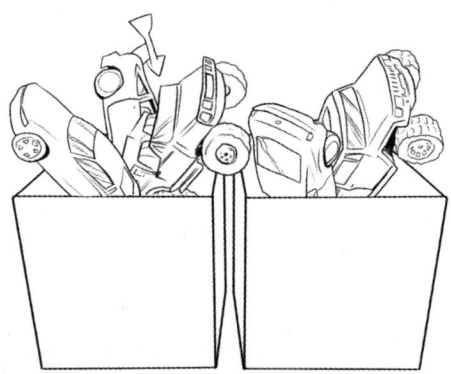

Most preschoolers can easily tell which box has more cars and which plate has more cookies.

But comparing larger quantities (or quantities are that look about equal) is much more difficult. To learn to make these more difficult comparisons, preschoolers need instruction and lots of practice. For example, in the picture below, most young children would find it very difficult to tell which bag has more marbles.

Preschoolers who are just learning to compare larger quantities begin by matching the objects one-by-one to see which group has more.

Then, as they gain more experience with comparing, children learn that they can use counting to compare quantities: "There are seven striped marbles and eight plain marbles. Eight comes after seven, so eight marbles is more than seven marbles." In Chapter 6, you will help your child learn to compare quantities and written numerals.

Most preschoolers find it difficult to tell which bag has more marbles.

Joining and taking away: Beginning addition and subtraction

Counting, subitizing, recognizing written numerals, and comparing all help children understand what numbers are. But preschoolers also need to begin to learn what they can *do* with numbers. That doesn't mean that you should make your child sit down and solve pages of addition and subtraction problems, though! What's most important at this stage is that your child understands the *meanings* of the operations: addition as joining two sets together,

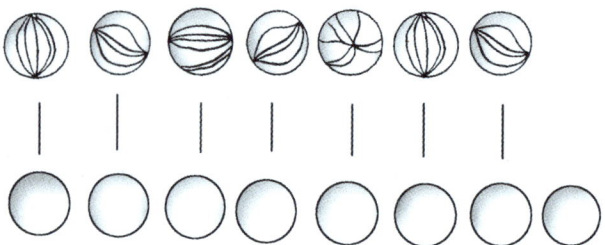

Matching one-by-one to compare two groups

and subtraction as taking part of a set away. In Chapter 7, you will use simple stories and concrete objects to introduce your child to the concepts of addition and subtraction.

How to Use This Book

Start when your child is interested

Most children will be ready to start the activities in this book around age four. If your child becomes interested in numbers at age three, go ahead and start trying some of the activities. No matter what age your child is, if there's frustration or tears, stop! Put the book away for a month and then give it another try. Sometimes, children's brains just need a little more time to mature before they're ready to understand number concepts.

Do the activities in order, but follow your child's lead for pacing

The activities in this book build math skills step-by-step. Each activity builds on the activities that came before, and each chapter builds on the skills developed in previous chapters. Plan to teach the activities in order without skipping around. (A few activities are labeled as optional, either because they are advanced for most preschoolers or because they require a lot of props. Feel free to skip those if they are too challenging for your child or too time-intensive for you to set up.)

Take as much time as you need with each activity. Generally, you'll want to spend a few (short!) teaching sessions on each activity so that your child understands it well before moving on to the next activity. Follow your child's lead. If your child has a lot of experience with a topic, you may breeze through several activities in one session. On the other hand, if your child struggles with a brand-new concept, you may want to stick with the same activity for a couple of weeks.

At the end of each chapter, you'll find a brief description of what your child should be able to do before moving on to the next chapter. Use this to help you decide whether to move on to the next chapter or give your child more practice with the current chapter.

Have a consistent (and short) math time

You're more likely to remember to do math if you choose a consistent time each day. You might do a little math with your preschooler before bedtime, or you might play with numbers together after morning snack each day.

Most four-year-olds have a very short attention span, so don't expect a focused, twenty-minute lesson. You'll be amazed at what your child will learn in a year if you do math together for just five minutes per day, several times per week.

Ask, "How many?"

As your child masters counting, she needs to learn that the last number she says when counting is the total number. For example, she may be able to point to blocks and *count* "one, two, three," but she may not be able to answer afterward when you ask *how many* blocks there are. Until she masters this important skill, make sure to ask her how many objects there are every time she counts. And, make sure you explicitly say how many objects there are every time you model counting.

Weave math throughout the day

Math is more than just numbers, but the other math concepts your preschooler needs to know will come up naturally during the day. Keep an eye out for opportunities to talk about patterns, clocks, shapes, and measurements, and use these teachable moments to introduce these ideas to your preschooler. This can be as simple as checking the time on the clock or measuring flour for pancakes together.

Have fun!

Math time for preschoolers should be *fun*! Feel free to adapt the games and activities to your preschooler's personality. Use her favorite trucks for the counting activities, or pretend that the counters on the ten-frame are people sitting on a bus.

Preschoolers especially love peek-a-boo games and being in charge. In this book, you'll see lots of hiding games and chances for your child to "be the teacher." Go ahead and make the hiding games as dramatic and fun as possible, and enjoy the chance for your child to show off what she knows when she leads activities. Most of all, enjoy this time with your preschooler.

What You Will Need

The activities in this book require only simple items from around your house, and many of the activities do not require any materials at all. However, it's much easier to do math consistently if you don't have to hunt for supplies. To make it as easy as possible to have math time each day, put together a small "Math Basket" with the most frequently-used materials, including:

- This book
- 20 small counters (You can use whatever you have around the house, such as pennies, Legos, or plastic cubes. It's best to use something simple and geometric so that your child will focus on counting, not examining the items. If your child still tends to put things in her mouth, make sure to use something large enough not to be a choking hazard, such as small wooden blocks.)
- Resources from the Appendix, preferably printed on cardstock (pages 89-101)
- One nickel and ten pennies
- Blank paper and a writing utensil

There are also some items that you will only need occasionally. If you would like to gather everything you'll need in advance, here is the full list:

- Ten small toys (like toy cars or small animal figures)
- Four stuffed animals
- Small blanket
- Two small paper bags
- Two regular, six-sided dice
- Tape
- Two different small objects for game tokens
- Two different-colored writing utensils
- Toy cups, plates, spoons, and play food items (optional)

- Dot stickers (optional)
- Small food items (raisins, pieces of cereal, etc.) (optional)
- Large index cards (optional)
- Small stickers (optional)

— CHAPTER 1 —
COUNTING TO FIVE

Chapter Overview

In Chapter 1, your child will learn the basic principles of counting as he learns to count to five. (See page 12 in the introduction for a list of these principles.) You'll also introduce your child to the concept of zero.

If your child already has a lot of experience with counting, you may be able to move through this chapter's activities quickly. But, as discussed in the introduction, there are many subtleties to counting that your child may not have learned yet, so don't skip the chapter entirely.

1.1   Count Five Toys in a Line

Purpose

Practice counting items one-by-one; practice counting to five

Materials

Five small toys, such as cars or plastic animals

Activity

Line up the five toys. Slowly and deliberately point to each animal and count it: "One...two...three...four...five! There are five toys." Ask your child to count the toys. Make sure he points to only one toy for each counting number. After he has counted, ask how many toys there are. (Five.)

Note

Children need lots of experience with counting to become proficient. If your child doesn't have much counting experi‐ ence, you'll want to prac‐ tice this activity with  other objects around the house. At first, continue to line up the toys so that it's easier for your child to keep track of which ones have already been counted.

Keeping track of which items have been counted also becomes more complicated when there are more items. For now, count groups of five objects or fewer so that your child achieves full mastery up to five. Your child will learn to count to ten in Chapter 2.

1.2   Count Toys in Different Arrangements

Purpose

Develop understanding that rearranging items doesn't change their total number; practice counting to five

Materials

Five small toys, such as cars or plastic animals

Activity

Line up the five toys. Ask your child to count the toys. Make sure she points to only one toy for each counting number. After she has counted, ask how many toys there are. (Five.) Then, spread the toys out in a random arrangement and ask your child to count them again. (Five.) Last, place the toys close together and ask your child to count them one more time. (Five.)

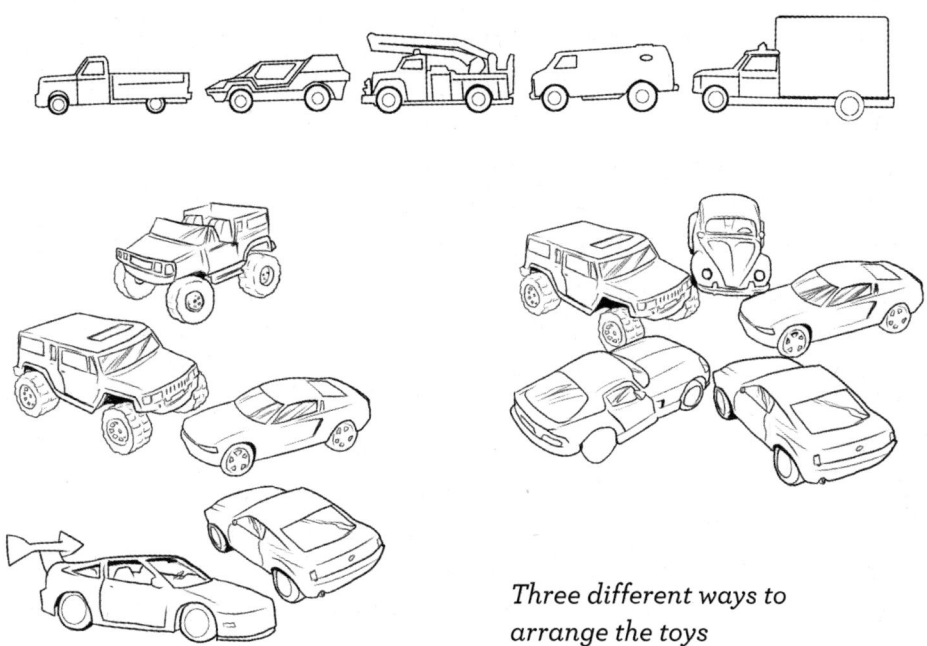

Three different ways to arrange the toys

Note

When you rearrange the toys, your child may simply tell you that there are five toys without counting them again. This shows that your child already understands that the number of objects does not depend on how they are arranged. But many children will continue to count all five objects as though each new arrangement might have a different number of objects in it. Repeated experience with counting will eventually convince your child that rearranging does not change the total number.

1.3 Count with Fingers to Five

Purpose

Practice counting to five with fingers

Materials

No special materials needed

Activity

Hold up one finger. Ask your child how many fingers you are holding up. (One.) Encourage him to copy you by holding up one finger. Then, hold up two fingers. Use your other hand to model how to count the two fingers one-by-one. Again, have your child copy you by holding up two fingers. Repeat this activity with three, four, and five fingers.

Note

Since our fingers are always available, they are one of the most natural and important counting "tools" for children. They're also great for practicing math when you're on the

2 fingers raised on a hand

go. Try doing this activity (or any of the finger games in this book) any time you need to keep your child occupied, such as at the grocery checkout or doctor's waiting room.

1.4   Introduce Zero with Fingers

Purpose

Introduce zero

Materials

No special materials needed

Activity

Review Activity 1.3 (Count with Fingers to Five). Then, hold up a closed fist with no fingers showing. Tell your child that we say there are zero fingers when there are no fingers showing. Ask your child to show zero fingers as well.

Zero fingers

1.5   Count Around the House

Purpose

Practice counting to five

Materials

No special materials needed

Activity

Walk around your home with your child. Ask "How many?" questions about the groups of objects you see. (Only ask about groups of five or fewer items.) For example:

· How many chairs are at the kitchen table?
· How many glasses are in the sink?
· How many books are on the nightstand?
· How many pencils are in the cup?

Also ask some silly questions whose answer is zero:

· How many elephants are in the living room?
· How many fairies are in the bathtub?
· How many spaceships are on the table?

You can also do this activity at a park, store, or backyard.

1.6   Match My Mat

Purpose

Practice matching items one-to-one; practice counting to five

Materials

Two pieces of paper; ten identical counters

Activity

Place four counters on a piece of paper, as shown. Ask your child to create a matching arrangement on the other piece of paper. Then ask your child how many counters there are on his paper. (Four.)

Repeat this activity with other numbers of counters up to five. Some possible arrangements include:

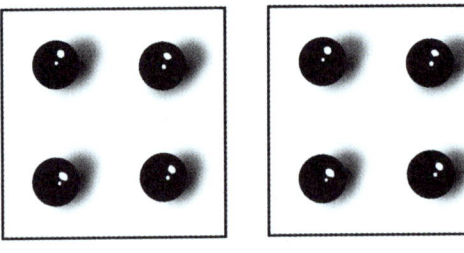

Your mat.   *Your child's mat.*

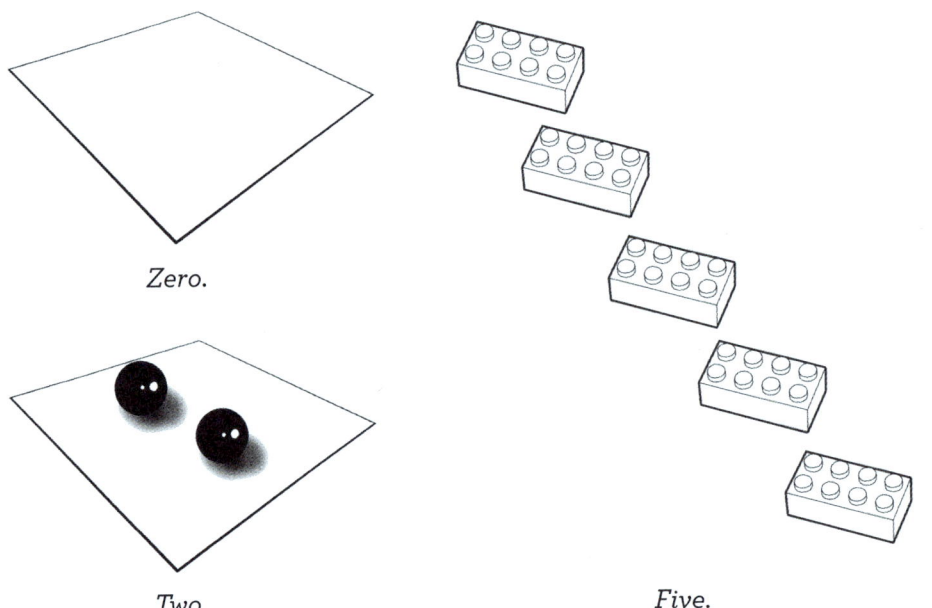

Zero.

Two.   *Five.*

Note

Understanding that we say one (and only one) word in the counting sequence for each object counted is essential for correct counting. In this activity, matching each of the counters on your paper one-by-one helps your child practice this important skill.

1.7   Set the Table for a Tea Party (Optional)

Purpose

Practice matching items one-to-one; practice counting to four

Materials

Four stuffed animals; four each of toy cups, plates, spoons, and play food items

Activity

Set the four stuffed animals on the floor (or on a picnic blanket or around a small table). Ask your child to help you prepare a tea party for the animals. First, ask your child to count the number of guests at the party. (Four.) Ask how many plates you will need. (Four.) Have your child count out four plates and put one in front of each stuffed animal. Do the same with the toy cups, spoons, and play food items.

If your child enjoys this activity, you can repeat it with other numbers of stuffed animals and different place settings. Your child can also practice one-by-one matching as he helps set the table for dinner.

Note

While it's obvious to adults that four stuffed animals require four plates, this is a concept that takes time to develop. If your child isn't sure how many plates are needed, have him place one plate in front of each stuffed animal and then count how many he used. With experience, he'll realize the connection between the number of guests and plates.

1.8   Feed the Stuffed Animal

Purpose

Learn to count out up to five objects when given a target number

Materials

Five counters; stuffed animal; blank piece of paper or toy plate

Activity

Set the blank paper or plate in front of the stuffed animal. Tell your child that the animal is hungry for two "cookies" (or apples) and ask her to put two of the counters on the plate for the animal to eat. For added fun, use the stuffed animal as a puppet and have the animal "eat" the plateful.

Repeat with other numbers of cookies up to five, including zero. (Perhaps the stuffed animal has a stomachache and doesn't want any more!) Then, let your child choose the number of cookies and check whether you've placed the correct number on the plate.

Note

It's often more difficult for children to count *out* a given number of counters than to count an already-existing set of counters. That's because the child has to keep the target number in mind and stop after each object to decide whether he's reached the target number. For more practice, look for opportunities during the day for your child to count out small sets of objects. For example, you might ask your child to hand you three forks or pick out two books to read before bedtime.

1.9   Count Hops

Purpose

Understand that intangible quantities like motions can be counted; practice counting to five

Materials

No special materials needed

Activity

Hop one time. Ask your child how many times you hopped. (One). Repeat with other numbers up to five. You can include zero by simply standing still and looking at your child expectantly.

Also reverse the process and give your child a target number of times to hop. You can vary the activity by using jumping jacks, somersaults, or whatever physical movement your child likes to do. Most preschoolers find it hilarious to hold perfectly still while doing "zero somersaults."

Note

Preschoolers often find it more difficult to count motions than objects. With objects, the entire set can be seen at once. But with motions, each movement is over before the next begins. Encourage your child to count each hop as you do it, so as not to lose track.

1.10   Match Claps to Counters

Purpose

Practice matching sounds and concrete objects one-to-one; practice counting to five

Materials

Five counters

Activity

Tell your child that you are going to clap a few times, and that you want him to put out one counter for each clap. Clap three times, slowly and deliberately without saying anything aloud. Encourage him to put out a counter after each clap, for a total of three counters. After you finish clapping, ask how many counters he put out and how many times you clapped.

Repeat for the other numbers up to five, and include zero claps, too.

Note

Counting sounds is even more abstract than counting motions because sounds aren't visible. This activity connects the concept of counting sounds with the already-familiar idea of counting concrete objects.

1.11   Count Claps

Purpose

Understand that intangible quantities like sounds can be counted; practice counting to five

Materials

No special materials needed

Activity

Slowly and deliberately clap two times. Ask your child how many times you clapped. (Two.)

Repeat with the other numbers up to five, including zero. Then reverse the process. Ask your child to clap three times. He may do this with or without counting each clap aloud, whichever makes more sense to him. Repeat with the other numbers up to five. You can also do this activity with stomps, taps on the table, or percussion instruments.

Is My Child Ready to Move On?

Your child is ready to move on to Chapter 2 when he can:

- Quickly and confidently count up to five objects and tell how many are in the set
- Count out a given number of objects (up to five)
- Count up to five sounds or motions and tell how many sounds or motions there were.

It's not expected that your child will fully understand the concept of zero yet. There will be more opportunities to learn about zero in the following chapters.

If your child is not quite ready to move on to Chapter 2, go back through the activities in Chapter 1 and have him practice counting to five until it is fully mastered. You can give the activities more variety by changing the toys and counters used for counting, or by counting items in different locations around or outside your home.

— CHAPTER 2 —
COUNTING TO TEN

Chapter Overview

In Chapter 2, your child will learn to count to ten. As in Chapter 1, she will learn the counting words first (this time, to ten) and practice counting objects in a line one-by-one. Then, she will learn to count up to ten scattered objects. Children often find it difficult to count larger sets of objects, because it becomes harder for them to keep track of which objects have and have not been counted. Through progressively more challenging counting activities, your child will learn how to count these larger sets systematically. She will also learn how to count out up to ten objects from a larger pile.

2.1 Hide and Seek to Ten

Purpose

Become familiar with the counting sequence to ten

Materials

No special materials needed

Activity

To help your child become familiar with the counting words to ten, play hide-and-seek. Close your eyes and count to ten, clearly and slowly. While you count, your child should hide somewhere in your home. (You may want to set some limits on possible hiding places before you begin.) After you reach ten, go and find your child. Play several times.

To prepare your child for being the seeker, have her count to ten in unison with you several times. Then, reverse roles and hide while your child counts to ten with her eyes closed. Play several more times with your child as seeker.

Note

While you are hiding, listen closely to your child's counting to see if she is saying the counting words correctly. If she makes mistakes, have her practice counting to ten with you before the next round of the game.

2.2   Count Ten Toys in a Line

Purpose

Practice counting items one-by-one; practice counting to ten

Materials

Ten small toys, such as cars or plastic animals

Activity

Line up the ten toys. Leave a gap between the two groups of five, as shown below. Slowly and deliberately point to each toy and count it: "One...two... three...four...five. Six...seven...eight...nine...ten! There are ten toys." Ask your child to count the toys. Make sure she points to only one toy for each counting number. After she has counted, ask how many toys there are. (Ten.)

Repeat this activity with six, seven, eight, and nine toys in a line. Continue to leave a gap between the first five toys and the rest.

Note

Leaving a space between the first five toys and the remaining toys (and pausing between five and six while counting) previews the idea that the numbers from six to ten can be thought of as combinations of five and some more. Your child will explore these combinations further in Chapter 4.

2.3 Count with Fingers to Ten

Purpose

Practice counting to ten with fingers

Materials

No special materials needed

Activity

Show your child how to count to ten on your fingers: slowly and deliberately count to ten, raising a new finger for each new number. Then, do it again. This time, have your child raise her fingers and count along with you.

Next, show eight fingers on your hands and ask your child to count how many fingers you are holding up. (Eight.)

Encourage her to count in a systematic way, starting at one side and counting all the fingers in order. Touching each finger as she counts it may help her keep track of which fingers she has already counted.

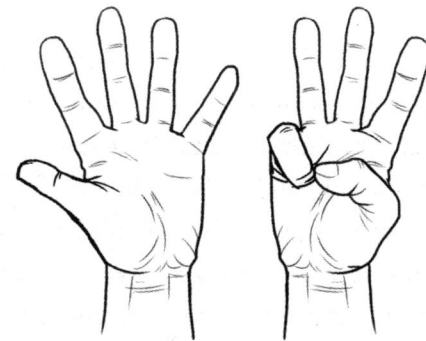

Repeat this activity for the other numbers of fingers from six to ten.

Note

When using your hands to show more than five fingers, it's best to hold up all the fingers on one hand before holding up any fingers on the other hand. Holding up the fingers in this way helps children begin to see how the numbers from six to ten can be thought of as five fingers plus some more.

2.4   Count Scattered Objects

Purpose

Learn to count objects systematically by moving each object into a pile after counting it; practice counting to ten

Materials

Ten counters

Activity

Scatter nine counters on the floor or table in front of your child and ask her to count them. Observe her as she counts. If she has trouble keeping track of which items she has already counted, suggest that she move the objects already counted to a separate pile. After she finishes counting, make sure to ask how many counters there are. (Nine.)

Repeat this activity with other numbers of counters from six to ten.

Note

As your child learned in Chapter 1, accurate counting requires counting each object once (and only once). Moving each object to a new pile after counting it makes it easier to keep track of which objects have been counted and which have not.

However, we often need to count things that cannot easily be moved from one pile to another, so the next two activities will help your child learn to count items *without* moving them. Make sure your child can confidently and accurately count items by moving them before moving on to Activity 2.5 ("Don't Move the Counters!").

2.5   Don't Move the Counters!

Purpose

Practice counting objects systematically without moving them; practice counting to ten

Materials

Ten counters

Activity

Scatter seven counters on the floor or table in front of your child. Pretend that the counters are glued to the table so that they can't be moved. (Dramatically pretending to lift a counter and failing will add to the effect.)

Ask your child to count the counters without moving them. If she needs help, suggest that she start in one corner of the arrangement and then count the counters systematically, either moving up and down or across the arrangement. Touching each counter as she counts it may also help her keep track.

Repeat this activity with other numbers of counters from six to ten.

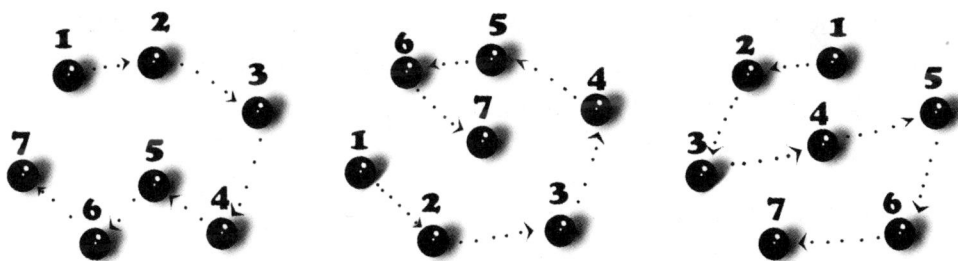

Possible ways for your child to count systematically

2.6   Count Pictures

Purpose

Practice counting pictures systematically; practice counting to ten

Materials

Five pieces of blank paper;
writing utensil; dot stickers
(optional)

Advance Preparation

To prepare for this activity, draw
the following patterns on the
pieces of paper.

There's no need to copy the
designs exactly, but try to scatter the dots randomly on the
paper. (Or, you can use dot stickers to make the designs instead
of drawing.)

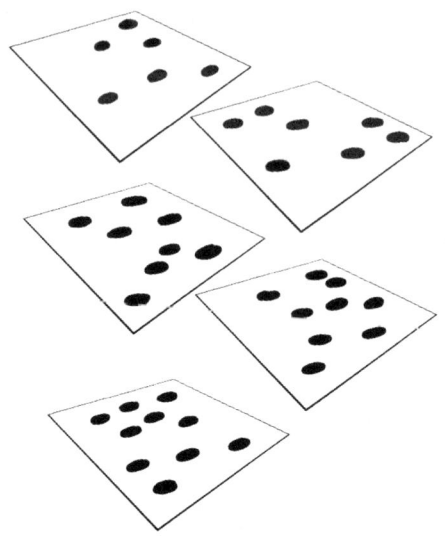

Activity

Show your child the paper with ten dots on it and ask her to count how
many dots there are. As in Activity 2.5 ("Don't Move the Counters!"),
encourage her to start in one corner of the arrangement and then count the
dots systematically, either moving up, down, or across the arrangement.
Touching each dot as she counts may also help her keep track. Make sure
she tells you how many dots there are after she finishes counting. (Ten.)

Repeat with the other dot designs.

2.7   Play Store with Pennies

Purpose

Learn to count out up to ten objects when given a target number

Materials

Ten pennies; a variety of small toys

Activity

Pretend that you are a storekeeper and that your child has come to shop
at your store. Give your child the ten pennies and display the small toys

as if they are for sale. When your child pretends to buy an item, name a "price" that is ten pennies or less. Have your child count out that number of pennies to pretend to pay you for the item.

For example, you might say, "Would you like to buy this toy train? It only costs eight pennies." Your child then counts out eight pennies and pretends to buy the train.

After your child pays for each item, give the pennies back. Play again for as long as your child is interested.

Note

As in Activity 1.8 (Feed the Stuffed Animal), this activity gives your child practice in counting out a quantity, not just counting an already-existing set.

2.8   Tens Treasure Hunt (Optional)

Purpose

Practice counting out ten items

Materials

No special materials needed

Activity

Announce that you and your child are going on a tens treasure hunt. Ask your child to find and count out some groups of ten items around your home. (If she's not sure where she can find ten objects, suggest that she look for small items that you have a lot of, like toothpicks, raisins, crayons, cotton balls, etc.)

As your child finds items, have her count out ten of the

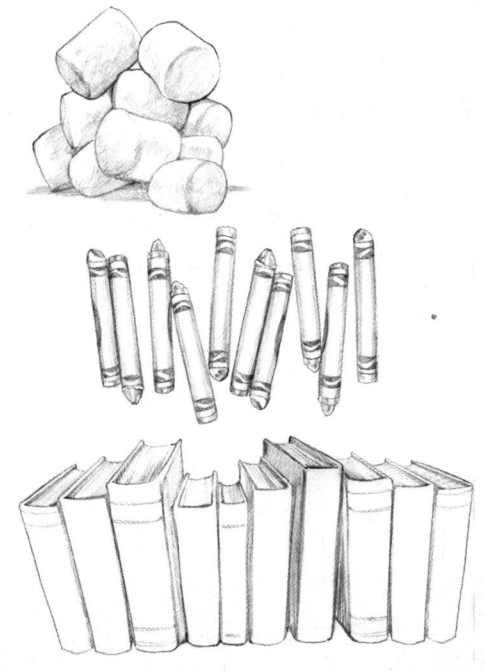

item and bring them to a central spot, like the kitchen counter. Then, challenge her to find a different group of ten. Continue gathering groups of ten for as long as she is interested.

When you've finished gathering the groups of ten, take a moment to look at all the piles together. Emphasize that all of the piles have ten objects, even though the piles are different sizes.

Note

Even though your child counted out the piles herself, she may have trouble believing that all the piles have the same number of objects. Young children tend to equate "bigness" with bigger numbers, so they expect that bigger piles must have a larger number of items. If that's the case for your child, keep the focus in this activity on counting out the tens. More experience with counting and size will help her realize that a pile of *three* beach balls can be larger than a pile of a *hundred* grains of rice.

Is My Child Ready to Move On?

Your child is ready to move on to Chapter 3 when she can:

- Count a set with up to ten objects confidently and systematically
- Count out ten objects from a larger pile.

If your child can't reliably count to ten yet, give her more practice with a greater variety of objects and incorporate counting as much as possible throughout the day. For example, count the steps every time you go upstairs, or look for pictures to count in your child's picture books.

— CHAPTER 3 —
NUMBERS FROM ZERO TO FIVE

Chapter Overview

In Chapter 3, your child will learn to recognize amounts up to five *without* counting. (This important skill is called subitizing; see pages 12-14 in the introduction for more explanation.) You'll use small groups of items around the house, your fingers, and a simple grid called a five-frame to help him learn to "see" at a glance how many objects there are in a group. Once your child has learned to recognize the quantities to five, he'll continue to use his fingers and the five-frame to learn the combinations that make five (four and one, two and three, five and zero).

3.1 I Spy Numbers

Purpose
Begin to think of quantities as groups rather than counting them one-by-one

Materials
No special materials needed

Activity
Secretly choose a set of two objects in the room you are in (for example, two pictures on the wall or two books on the coffee table), and say, "I spy with my little eye two of something." Encourage your child to guess your

secret objects. He might ask, "Are you thinking of those two toy cars on the rug? Or the two lamps by the window?" Try to choose objects that are large, obvious, and close together. If your child needs help, give clues about the color or size of the objects, or look directly at the objects so your child can follow your gaze.

Play several times, including sets with one, two, three, four, and five items. Then, reverse roles and let your child choose secret objects for you to guess.

Note

In Chapter 1, your child counted out small objects and counted items in your home. He may find looking for a set with a certain number of objects more challenging, because he has to think about what the number means, hold the number in his head, and look for the matching set.

Also note that zero isn't included in this activity, since it is very difficult to "spy" zero of an object!

3.2   Hide the Toys

Purpose

Begin to recognize quantities up to five without counting; explore the combinations that make five (five and zero, four and one, two and three)

Materials

Five small toys (such as cars or plastic animals); small blanket

Activity

Line up the five toys and ask your child how many there are. (Five.) With your child's eyes closed, secretly hide one toy under the blanket. When he opens his eyes, ask how many toys are still showing. (Four.) Also ask how many he thinks are under the blanket. (One.) If your child's not sure, have him look at how many lumps are under the blanket.

After he answers, let him look under the blanket to see if he was correct. Then, move the hidden toy back to the line and play some more, hiding different numbers of toys each time. To include zero, move all of the

toys under the blanket (so that five are under the blanket and zero are visible). Or, leave all the toys in the line, so that five are visible and zero are under the blanket.

Allow your child to hide the toys and ask you how many there are, too.

Lining up the toys (rather than scattering them randomly) makes it easier to tell how many are missing. When your child is ready for more of a challenge, arrange the five toys in a haphazard pattern before hiding any under the blanket.

Note

Many preschoolers can easily recognize groups of one, two, or three objects but have trouble distinguishing between four and five objects. One way to help your child tell the difference between four and five is to point out that a line of five always has one object in the middle with two objects on either side.

On the other hand, a line of four objects does not have a middle object.

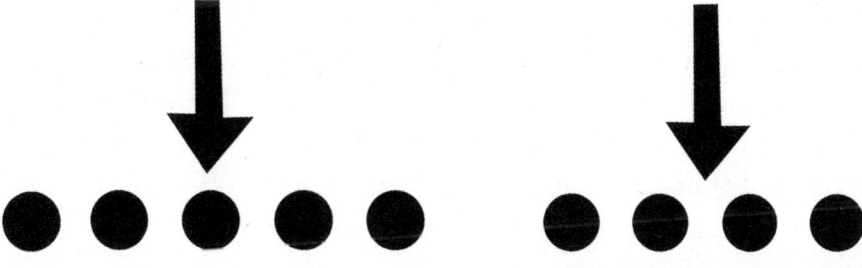

A line of five always has one counter in the middle, with two counters on either side.   *A line of four has a space in the middle, not an object.*

3.3　Peek-A-Boo Fingers

Purpose

Become faster at recognizing quantities up to five without counting

Materials

No special materials needed

Activity

Review Activity 1.3 ("Count with Fingers to Five"). Then, put your hand behind your back and hold up three fingers. Tell your child that you are going to show your fingers very quickly, and his job is to tell how many fingers you're holding up, as fast as he can.

With drama, briefly show your child your hand and then put it behind your back again. Ask him how many fingers you are holding up. (Three.) After he has guessed, bring your hand out again to let him count the fingers and see whether he was right.

Repeat with the other numbers up to five, including a closed fist for zero. Then, reverse roles. Let him hide his hand and ask *you* how many fingers he's holding up.

Note

Keep this activity (and all of the peek-a-boo activities in this book) fast-paced and fun, adjusting the speed to your child. Try to flash the fingers as quickly as possible so that your child doesn't have time to count one-by-one but instead begins to recognize the quantities. Never push your child to guess, though, and always allow your child to count one-by-one any time he's not sure or makes a mistake. With repeated practice, he'll learn to recognize quantities up to five instantly.

3.4   Fingers Up, Fingers Down

Purpose

Begin to identify combinations that make five (five and zero, four and one, two and three)

Materials

No special materials needed

Activity

Show your child four fingers, with your palm facing you. Ask how many fingers are up (four). Then, ask how many fingers are down without turning your hand around.

After your child responds (one), turn your hand so that your palm faces your child and he can confirm his answer.

Repeat with other numbers of fingers. Then, let your child be the teacher and quiz you on how many fingers he has up and down.

Note

It might take your child a minute to figure out how many fingers are down. Make sure to give him some thinking time before offering

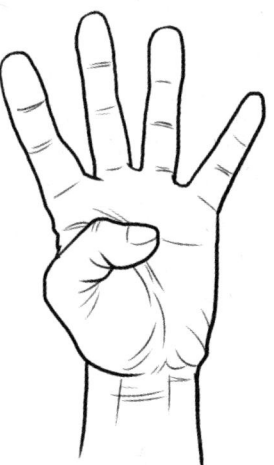

Four fingers up, one finger down

help. Children sometimes copy their parent's hand with their own hand to see how many are down, and sometimes they look closely at the edge of the parent's hand to see where the tucked fingers' edges are. Whatever method makes sense to your child is fine; what's most important is that he realizes he can reason out the answer. If he makes a mistake, just turn your hand around to show him how many fingers are down and then try again.

3.5   Counters on the Five-Frame

Purpose

Introduce the five-frame

Materials

Five-frame (page 89); five counters

Activity

Place three counters on the five-frame as shown below. When using the five-frame, always place the counters from left to right, with no empty spaces between counters.

Ask your child how many counters there are. (Three.) As in the other activities in this chapter, encourage him to recognize the quantity by sight, but allow him to count if he's not sure.

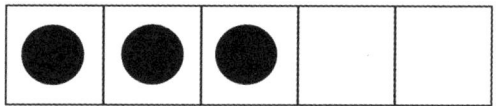

How to place three counters on the five-frame

Repeat for other numbers up to five, including zero. Then, tell your child a number from zero to five and have him put the matching number of counters on the five-frame.

Note

This activity introduces your child to the five-frame. This simple grid helps children visualize quantities in an orderly way. It also helps children begin to notice relationships between numbers, which is essential for comparing numbers and for beginning addition and subtraction. For example, when a child puts three counters on the five-frame, he can immediately see that he would need to add two more counters for all five boxes to be filled.

This builds a mental framework for comparing numbers (three is less than five) and and for addition (two

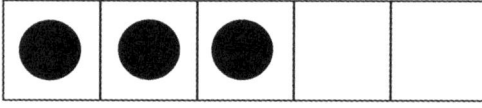

Three boxes are full, and two are empty.

plus three equals five). You'll use the five-frame (and its cousin, the ten-frame) more throughout this book.

3.6   Peek-a-Boo Counters on the Five-Frame

Purpose

Practice identifying quantities up to five without counting

Materials

Five-frame (page 89); five counters; blank piece of paper

Activity

Ask your child to close his eyes. While his eyes are closed, secretly place one counter on the five-frame and cover the five-frame with the blank piece of paper. Tell him to open his eyes, and then remove the piece of paper for a second and cover the five-frame again. Ask how many counters are on the frame. (One.) If he's not sure, remove the paper and let him look for as long as he needs.

Repeat with other numbers of counters from zero to five.

Note

As in Activity 3.4 (Fingers Up, Fingers Down), try to reveal the counters for just a second so that your child doesn't have time to count one-by-one but instead begins to recognize the quantities by sight. Adjust the pace to your child, going as fast as possible without frustrating him. Also, make sure your piece of paper isn't see-through. Construction paper or a manila file folder works perfectly.

3.7   Missing Counters on the Five-Frame

Purpose

Learn to visualize the combinations that make five (five and zero, four and one, two and three)

Materials

Five-frame (page 89); five counters; blank piece of paper

Activity

With your child watching, place four counters on the five-frame. Ask how many counters there are. (Four.) Then, ask how many *more* counters it would take to fill the five-frame. (One.)

Repeat with other numbers of counters from zero to five. To make the activity more challenging, you can make it a peek-a-boo game: cover the five-frame with a piece of paper and remove it briefly as in Activity 3.6 ("Peek-a-Boo Counters on the Five-Frame"). Then, ask how many counters are on the five-frame and how many more it would take to fill it.

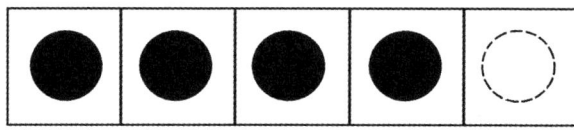

It takes one more counter to fill the five-frame.

3.8   Hidden Handfuls

Purpose

Practice identifying combinations that make five (five and zero, four and one, two and three)

Materials

Five   counters;   five-frame (page 89)

Activity

Show your child all five counters, then cover them with both your hands. Secretly put two counters under your right hand and three under your

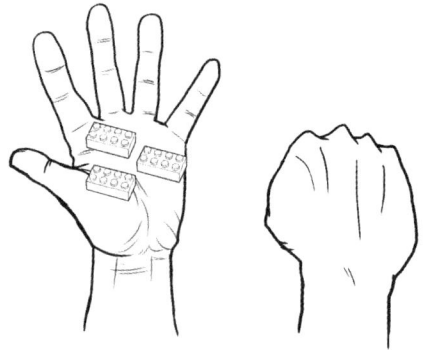

"If I started with five counters, how many must be in my right hand?"

left hand. Then, turn over your left hand to show the three counters. Ask how many must be in your right hand. (Two.)

If your child is not sure, put the three counters on the five-frame. Ask him to imagine all five counters on the five-frame and think about how many are missing. Since there are two empty boxes, there must be two counters hidden under your right hand.

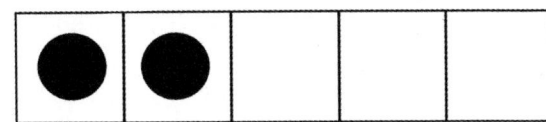

After your child answers, open up your right hand so he can check his answer. Repeat with other combinations of five counters, and let your child quiz you, too.

Is My Child Ready to Move On?

Your child is ready to move on to Chapter 4 when he can:

- Look at up to five objects in a line and immediately tell how many there are, without counting
- Identify combinations that make five, using his fingers or the ten-frame.

If your child is not yet able to identify groups up to five without counting, practice the activities in this chapter some more. Pay special attention to groups of four and five, encouraging your child to look and see whether there is an object in the middle of the line to help tell whether there are four or five objects.

NUMBERS FROM SIX TO TEN

Chapter Overview

In Chapter 4, your child will learn to recognize six to ten objects as combinations of "five and some more" (six is five and one more, seven is five and two more, and so on). You'll use your fingers, a simple grid called a ten-frame, and nickels and pennies to represent these combinations.

After your child has become familiar with thinking of the numbers from six to ten as "five and some more," she'll then learn the combinations that make ten (five and five, six and four, seven and three, eight and two, nine and one, ten and zero).

4.1 Five Fingers and Some More

Purpose

Introduce the idea that the numbers from six to ten can be thought of as five plus some more (six is five plus one, seven is five plus two, and so on)

Materials

No special materials needed

Activity

Briefly review Activity 2.3 ("Count with Fingers to Ten"): count to ten on your fingers, raising a new finger for each new number and encouraging your child to count along with you.

Then, show your child six fingers (five on one hand, and one on the other hand). Ask how many fingers you are holding up on each hand. (Five on one hand; one on the other.) Ask how many fingers you are holding up altogether on both hands. (Six.)

If your child immediately knows how many fingers you are holding up, ask her how she knows. (Many answers are possible, but most children will respond that they know six is one more than five.) If your child counts each finger one-by-one, point out that one hand has five fingers, so you can just count on from five: "This hand has five fingers up, so I can count one more past five. Five (wiggle hand with all fingers up)...six (wiggle single finger)."

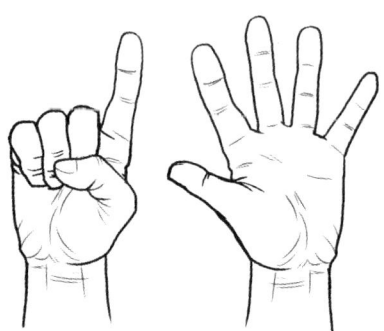

Repeat with seven, eight, nine, and ten fingers. Emphasize how much more each number is than five, and encourage your child to count on from five rather than counting each finger individually.

Note

When children begin counting from a number other than one, this is called *counting on*. For example, a first-grader adding nine and two might count on from nine: "Nine...ten...eleven!" In this activity, you encourage your child to count on when you suggest that she start counting at five to find how many fingers they are. It can take a long time for preschoolers to realize that they can count on rather than always starting at one when counting. Throughout this chapter, model counting on, but never push your child to use counting on if it doesn't make sense to her.

4.2 Peek-A-Boo Fingers with Two Hands

Purpose

Begin recognizing the numbers from six to ten as five plus some more (for example, six is five plus one, seven is five plus two, and so on)

Materials

No special materials needed

Activity

Put your hands behind your back and hold up seven fingers (five on one hand, two on the other). Briefly show your child your hands before putting them behind your back again. Ask her how many fingers you are holding up.

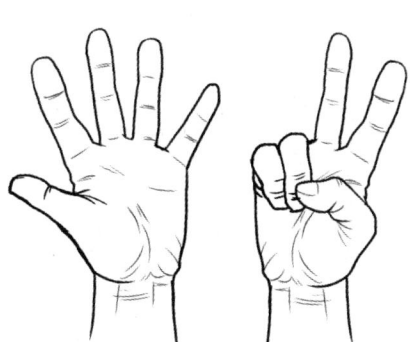

After she guesses, show your child your hands again and let her check her answer.

Repeat with the other numbers from six to ten. Also let your child hide her hands and ask *you* how many fingers she's holding up.

Note

Children typically go through three stages as they learn to recognize these combinations. In the first stage, they count all the fingers, starting at one. In the second stage, they count on from five. In the third stage, they recognize the total number of fingers immediately. Different children will need different amounts of time to progress through these three stages. If you've played this game a few times and your child is still counting all the fingers starting at one, go on to the other activities in the chapter and see if the other representations help your child begin to move to the "counting on" stage.

4.3   Race to Ten Game

Purpose

Introduce the ten-frame

Materials

Double ten-frames (page 101); coin with heads and tails; 20 counters

Activity

Before playing this game, show your child one of the ten-frames. Ask her to count how many boxes it has. (Ten.) Compare it to the five-frame and point out that each ten-frame has two groups of five put together, with a thick line separating the two groups of five.

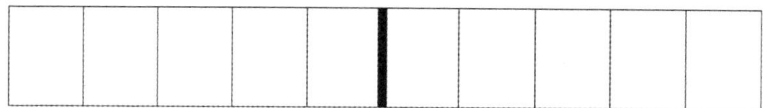

The object of this game is to fill in an entire ten-frame before the other player. To play, flip the coin on your turn. If it is heads, place one counter on your ten-frame. If it is tails, place two counters on your ten-frame. Say how many counters you have altogether. Then, it is the other player's turn. Continue until one person has filled in their entire ten-frame.

As the ten-frame begins to fill, encourage your child to notice how many counters more than five she has when she announces her total number of counters.

Note

As your child figures out her total number of counters after each play, continue to encourage her to count on from five rather than starting to count at one each time.

4.4 Hidden Counters on the Ten-Frame

Purpose

Practice counting on from five; become more familiar with thinking of the numbers from six to ten as "five and some more"

Materials

Single ten-frame (page 99); ten counters; blank piece of paper

Activity

Have your child watch as you place seven counters on the ten-frame. (Do not count them aloud or say how many there are.) Cover the first five counters with the blank piece of paper.

Ask your child how many counters are on the ten-frame. (Seven.) Encourage her to remember that there are five counters under the paper, so she can "count on" from five to figure out the total amount: "Five, six, seven!"

Repeat for the other numbers from six to ten.

If your child has trouble counting on from five, suggest that she imagine the five counters under the paper and count them in her head. If that is confusing for her, lift the paper and allow her to count them directly.

Note

If your child can readily tell how many counters are on the ten-frame *without* counting, you can make the activity more challenging by making this a peek-a-boo game. Secretly place counters on the ten-frame and cover the entire frame with a piece of paper. With your child looking, remove the piece of paper for two seconds and then cover the ten-frame again. Ask your child to tell how many counters were on the ten-frame.

4.5   Nickels and Pennies (Optional)

Purpose

Practice counting on from five

Materials

One nickel; five pennies; a variety of small toys (optional)

Activity

Show your child the nickel and pennies. Explain that each penny is worth one cent, but the nickel is worth five cents. Explain that if you were buying something that cost five cents, you could use five pennies, or just one nickel.

Show your child one nickel and one penny, and ask how much the coins are worth. (Six cents.) Encourage your child to think of the nickel as five cents and then count on to find that the two coins are worth a total of six cents.

Repeat for seven cents, eight cents, nine cents, and ten cents.

If you would like to give your child more practice with nickels and pennies, play store again as in Activity 2.7 ("Play Store with Pennies"). Lay out some small toys and pretend to be selling them. When your child pretends to buy an item, name a price from six to ten cents. Have your child use the nickel and pennies to count out the correct payment. Play again for as long as your child is interested.

Note

Using nickels requires fairly abstract reasoning from preschoolers, because they must understand that you can use *one* object to stand for *five* cents. This is different from the hidden counters on the ten-frame because the child knew that all five counters were really there, even if they were hidden by a piece of paper. If the idea of one nickel representing five cents goes over your child's head, feel free to skip this activity.

4.6 Fingers Up, Fingers Down to Ten

Purpose

Practice recognizing quantities from six to ten as five and some more; explore the combinations that make ten (five and five, six and four, seven and three, eight and two, nine and one)

Materials

No special materials needed

Activity

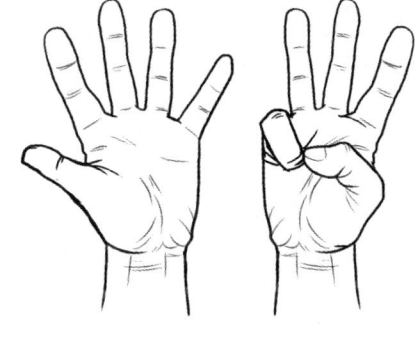

Show your child eight fingers, with your palms facing you. (As in previous activities, raise five fingers on one hand and three fingers on the other hand.) Ask how many fingers are up. (Eight.) Then, ask how many fingers are *down* without turning your hand around. (Two.)

After your child responds, turn your hands so that your palms face your child and she can see for herself how many fingers are down. Repeat with the other numbers of fingers from six to ten. (Make sure to include ten fingers so that your child can tell you that zero fingers are down.) Then, let your child be the teacher and quiz you on how many fingers she has up or down.

Note

Children usually find this activity easier when they start by seeing and identifying the larger number in each combination (for example, first identifying nine fingers up and then figuring out that one finger is down, rather than the other way around). If your child can easily tell you how many fingers are up and down when you're holding up six to ten fingers, you can make the activity more challenging by holding up smaller numbers of fingers and asking your child to say how many fingers are up and down.

4.7 Missing Counters on the Ten-Frame

Purpose

Practice recognizing quantities from six to ten as five and some more; explore the combinations that make ten (five and five, six and four, seven and three, eight and two, nine and one)

Materials

Single ten-frame (page 99); Ten counters; blank piece of paper

Activity

Place nine counters on the ten-frame and ask how many counters there are. (Nine). Ask how many more counters it would take to fill the ten-frame. (One.)

Repeat with the other numbers from six to ten. Then, make the activity a peek-a-boo game: cover the ten-frame with the paper, remove it briefly, and ask how many counters are on the ten-frame and how many more it would take to fill it.

Note

As in Activity 4.6 ("Fingers Up, Fingers Down to Ten"), if your child can easily tell you how many boxes are full and empty when there are six to ten counters on the ten-frame, you can make the activity more challenging by placing smaller numbers of counters on the ten-frame and asking your child to say how many boxes are full and empty.

Is My Child Ready to Move On?

Your child is ready to move on to Chapter 5 when she can:

- Look at up to ten counters on the ten-frame and tell how many there are, without counting every object one-by-one. (It's fine if she counts on from five.)
- Identify combinations that make ten, using her fingers or the ten-frame.

This chapter is meant to *introduce* your child to the combinations of "five and some more" and the combinations that make ten. Your child will learn more about these combinations in kindergarten; she does not need to memorize them before moving on to the next chapter.

— CHAPTER 5 —

WRITTEN NUMERALS FROM 0 TO 10

Chapter Overview

In Chapter 5, your child will learn that written numerals can be used to represent numbers, and he will learn to recognize the written numerals from 0 to 10.

    To help your child master the written numerals, you'll use activities that explicitly connect the written numerals with the quantities and spoken numbers that he is already familiar with.

Three ways to represent the number 4

5.1 Number Snack

Purpose
Introduce the written numerals from 0 to 10; understand that written numerals represent quantities

Materials
11 large index cards or half-sheets of paper; writing utensil; 55 small food items (raisins, pieces of cereal, etc.) (If you prefer not to use food items, you can use counters instead.)

Activity

Spread the index cards in a horizontal line in front of you and your child. With your child watching, write "0" at the bottom of the first card, "1" on the next card, and so on up to "10."

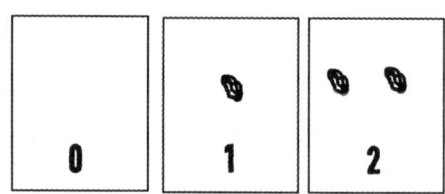

Then, have your child count out and place a matching number of snacks on each card (for example, zero raisins on the card labeled "0," one raisin on the card labeled "1," and so on up to ten raisins).

Ask your child to count from zero to ten, pointing to each matching card as he counts. Then, enjoy the snack together!

Note

Introducing the written numerals by making sets of concrete items helps make it clear that the written numerals are simply another way of representing quantities.

Save these cards for Activity 5.2 ("Making Number Pictures).

5.2   Making Number Pictures

Purpose

Become more familiar with written numerals from 0 to 10; understand that written numerals represent quantities

Materials

11 number cards from Activity 5.1 ("Number Snack"); writing utensil or 55 small stickers

Activity

Spread the cards from Activity 5.1 ("Number Snack") in order in a horizontal line. To review the numerals, ask your child to count from zero to ten and point to each matching card as he counts.

Then, work with your child to draw a matching number of small, simple pictures on each card. For example, you might draw nothing on the card labeled "0," one square on the card labeled "1," two suns on the card labeled "2," and so on.

Note

Preschoolers vary in how much they enjoy drawing (and how much stamina they have). If your child enjoys drawing, he might prefer to do all the drawings himself. If that's the case, you may need to spread out the project over a couple days so as not to wear him out with too much writing.

If your child doesn't like drawing, work together on the cards and take

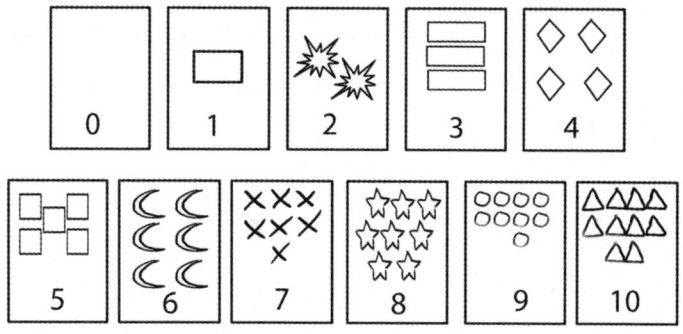

Sample set of cards.

turns drawing pictures. Or, use small stickers instead and have your child place the matching number of stickers on each card.

5.3 Number Race Game

Purpose

Practice naming written numerals from 0 to 10

Materials

Number Race game board (page 91); coin with heads and tails; two different small objects for game tokens

Activity

To set up the game, put both tokens on the space marked "0". On your turn, flip the coin. If the coin shows heads, move forward one space. If the coin shows tails, move forward two spaces. Say both the number you start on and the number(s) you land on. (For example, if you start on three and flip tails, say, "three, four, five," as you move your token.) Then, it is the other player's turn. Continue until one person has landed exactly on ten.

Note

This is a short, fast-paced game, so play several times. Saying both the number you start on and the numbers you land on reinforces the connection between the spoken numbers and written numerals, as well as the order of the numbers.

5.4　Number Jump

Purpose

Practice naming written numerals

Materials

11 sheets of blank paper; writing utensil; tape

Activity

With your child, write each numeral from 0 to 10 on its own sheet of paper. Make the numbers large and easy to see. Have your child help you lay the papers on the floor in order from 0 to 10. (You may need to use a little tape to keep them from sliding.) Then, have your child start at 0 and jump on the papers in order up to 10, shouting out each number as she jumps onto it.

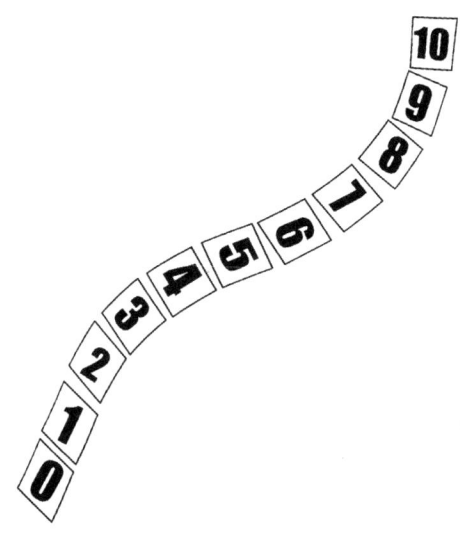

Repeat the activity a few times. To vary the activity, your child might use different movements each time (for example, hopping like a frog to the next paper, or hopping on one foot).

Note

As in Activity 5.3 ("Number Race Game"), your child's knowledge of the counting sequence helps scaffold this activity for him. For example, a child might not be quite sure yet what "7" means, but if he sees that it's on the next paper after six, he knows it must be seven.

Save these papers for Activity 5.5 ("Mixed-Up Number Jump").

5.5 Mixed-Up Number Jump

Purpose

Practice matching spoken numbers with written numerals

Materials

11 sheets of paper from Activity 5.4 ("Number Jump"); tape

Activity

Lay the numbered sheets of paper on the floor. Spread them out in random order, and use a little tape to keep them from sliding when jumped on, if needed.

Start by having your child stand on 0. Then, have him run to 1, then to 2, and so on up to 10. Once he knows where each paper is, challenge him to run from 0 to 10 as fast as possible.

You can also vary the activity by starting with your child in the middle of the room. Shout out a number from zero to ten and have your child run to that number as quickly as possible. Continue with the rest of the numbers from zero to ten in random order.

Note

Spreading the papers out in random order makes this activity more challenging than Activity 5.4 ("Number Jump"), where the papers were laid out in numerical order.

5.6 Number Card Matching

Purpose

Practice matching written numerals

Materials

Two sets of number cards 0–10 (pages 93-97)

Activity

Lay out one set of number cards in order from 0 to 10 horizontally. Shuffle the other set of number cards and place them in a face-down pile. Turn the top card face up. Have

0	1	2	3	4	5	6	7	8	9	10
2										

your child name the number and lay the card directly below its match. Continue with the rest of the cards.

Once the cards are all laid out, have your child point to each card and name the numeral, starting at 0 and continuing to 10.

5.7 Number Card Line-Up

Purpose

Practice naming the written numerals and putting them in order

Materials

One set of number cards 0–10 (pages 93-97)

Activity

Shuffle the number cards and lay them all out, face up. Ask your child to put the cards in order from zero to ten.

After he has put the cards in order, shuffle the cards again and challenge him to put them in order again as quickly as possible.

Note

This activity is quite a bit more difficult than the previous activities in the chapter because your child no longer has an existing visual model to match. If this activity is very difficult for your child, you may want to review and practice some of the earlier activities in the

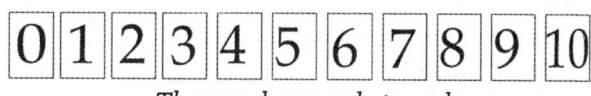

The number cards in order

chapter before moving on to Activity 5.8 ("Number Memory Game").

5.8   Number Memory Game

Purpose

Practice matching and naming written numerals

Materials

Two sets of number cards 0–10 (pages 93-97)

Activity

Shuffle both sets of number cards together and lay them out in a grid as in the game Concentration. On your turn, flip over two cards. If the two cards match, keep the pair. If not, turn them back over.

Take turns until all cards have been matched. Whoever finds the most pairs wins.

Note

Most preschoolers love this classic game. Encourage your child to say the numbers aloud as he finds them (and say them aloud yourself, too) to reinforce the connection between the spoken numbers and written numerals.

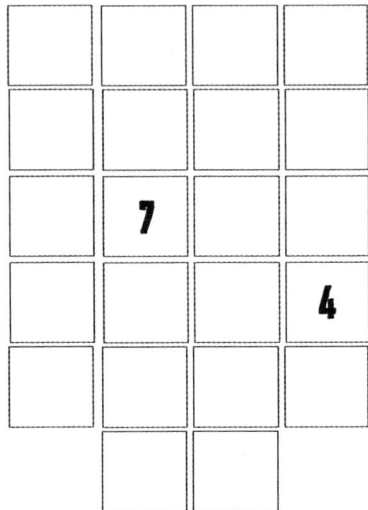

5.9   Go Fish Game

Purpose

Practice matching and naming written numerals

Materials

Two sets of number cards 0–10 (pages 93-97)

Activity

Shuffle both sets of number cards together, and deal out five cards to each player. Spread the rest of the cards out face-down to be the "fish pond." On your turn, ask the other player for the match to one of the cards you have in your hand. He must give you the card if he has it. If he doesn't have the card, he says, "Go fish!" and you take a card from the fish pond. If you get a match, place the matching cards on the table in front of you. Continue until all cards have been matched.

5.10   Number Scavenger Hunt (Optional)

Purpose

Match written numerals to real quantities

Materials

One set of number cards 0–10 (pages 93-95)

Advance Preparation

In this activity, your child will look for groups of objects around the house. To prepare, make sure that there is a group of objects in your home that matches each of the numbers from zero to ten. (Of course, for zero, you don't need any objects at all!) You likely already have plenty of groups with one to five objects (four chairs around the table, three apples in the bowl, etc.), but you may need to create groups for some of the larger numbers. You might leave a pile of seven books on the coffee table, place a box of eight crayons on a shelf, line up nine toy cars in a row, or stack ten blocks into a tower.

Activity

Tell your child that you've created a scavenger hunt for him. Give him the set of number cards and ask him to find a group of objects to match each of the number cards. Once he finds a match, he should place the number card by the group of objects.

After he's found groups to match all the number cards, return to each of the cards. Together, count the items in the group to confirm that the quantity matches the written numeral on the card. Accept any group that matches the number, even if it's not the one you were intending when you set up the scavenger hunt.

Is My Child Ready to Move On?

Your child is ready to move on to Chapter 6 when he can readily identify the written numerals from 0 to 10. If he needs more practice, go back and play some of his favorite games from this chapter.

— CHAPTER 6 —

COMPARING QUANTITIES AND NUMBERS

Chapter Overview

In Chapter 6, your child will learn to understand the concept of comparison (and the words *more*, *fewer*, and *equal*) by comparing groups of concrete objects. At first, she will compare quantities visually, but then she will use her counting skills to compare quantities. She'll also represent written numerals with concrete objects to help her begin to compare written numerals rather than groups of objects.

6.1 Compare Cookies, Part One

Purpose

Introduce the concept of comparing quantities

Materials

20 counters; two stuffed animals; two blank pieces of paper or toy plates

Activity

Set a plate or blank piece of paper in front of each stuffed animal. Pretend that the counters are cookies and that you are serving up a treat for each animal. Put a large handful of counters on one plate (about eight, but don't count) and put just two counters on the other plate.

Ask which animal has more cookies. (The one with the large handful of counters.) Ask which animal has fewer cookies. (The one with just two counters.) Encourage your child to answer these questions in full sentences. For example: "The teddy bear has fewer cookies than the bunny. The bunny has more cookies than the teddy bear."

Repeat with other amounts of cookies on each plate. Continue to make one amount much larger than the other so that it is easy for your child to tell at a glance which plate has more cookies. (For example, you might put seven counters on one plate and three on the other, or ten counters on one plate and five on the other.) Include zero cookies by occasionally leaving one of the plates empty.

Note

Make sure that all of the counters are the same size so that each takes up the same amount of space. Preschoolers find the concept of more and fewer very confusing when the objects are not the same size. For example, when asked to compare three beach balls with ten golf balls, some preschoolers will say that there are more beach balls, because the beach balls take up more space. In this chapter, to keep the focus on numbers, make sure to use objects that are all the same size.

6.2   Compare Cookies, Part Two

Purpose

Compare small quantities (for example, two counters versus four counters)

Materials

20 counters; two stuffed animals; two blank pieces of paper or toy plates

Activity

As in Activity 6.1 ("Compare Cookies, Part 1"), set a plate or blank piece of paper in front of each stuffed animal. Again pretend that the counters are cookies and that you are serving up a treat for each animal. Serve one animal four cookies and the other animal two cookies, then ask your child how many cookies each animal has. (Four and two.) Next, ask which animal has more. (The animal with four cookies.) Also ask which animal has fewer cookies. (The animal with two cookies.)

Continue with other amounts five or less. Sometimes give both animals the same number of cookies, and encourage your child to say that both animals have the same amount. Also make sure to include zero.

If your child is ever unsure about which animal has more, have her take the cookies off the plate and line them up next to each other. Make sure she keeps the spacing even and matches the cookies one-to-one to tell which line of cookies has more.

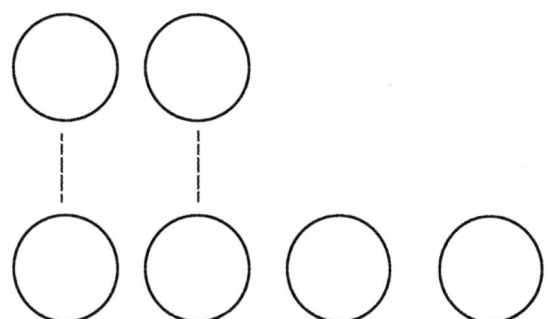

Comparing two and four counters by matching one-to-one

Note

Naming the number of cookies before comparing helps children begin to use their knowledge of numbers to compare quantities rather than just looking to see how much space the objects take up.

6.3 Equal Fish in the Pond

Purpose

Introduce the word *equal*; create equal sets

Materials

20 counters; blank paper; writing utensil

Activity

On a piece of paper draw two large circles. Tell your child that each circle is a fish pond. Pretend that the counters are fish and place four counters in one of the ponds. Then, ask your child to put the same number of fish in the other pond. After she has put four counters in the other pond, tell her that the two ponds have an *equal* number of fish.

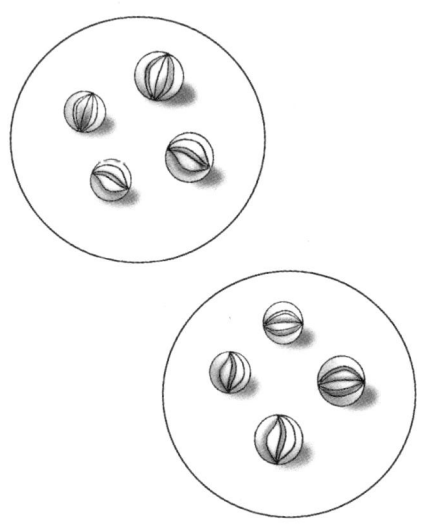

Then, clear the piece of paper. Repeat the activity with six, three, zero (an empty paper), and eight. Make sure to use the word *equal* as you discuss each arrangement.

Note

The different numbers used in this activity require different kinds of thinking from preschoolers. Most will be able to "see" when the two sets of four fish match each other. But to match six or eight fish, your child will need to carefully count both the set that you create and the set that she places in her fish pond. Learning to count while comparing helps your child move towards eventually comparing numbers without first representing them with concrete objects.

6.4 More and Fewer Fish in the Pond

Purpose

Create sets with more or fewer items than a given set

Materials

20 counters; blank paper; writing utensil

Activity

Begin by reviewing Activity 6.3 ("Equal Fish in the Pond"). Draw two large circles to be fish ponds. With your child watching, place five counters in one of the circles. Pretend that the counters are fish and ask how many fish are in your pond. (Five.) Ask your child to put an equal number of fish in the other pond. Then, add one counter to your pond. Ask which pond has more fish now. (Your pond.) Ask which has fewer fish. (Your child's pond.)

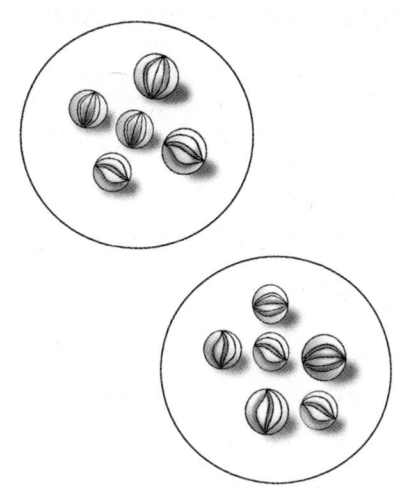

Clear the counters off the paper. Next, put three fish in your pond. Ask your child to put more fish in her pond than are in your pond. (Any number more than three is fine.) Repeat for two, five, and eight fish.

Then, place six fish in your pond. Ask your child to place fewer fish in her pond than are in your pond. (Any number less than six is fine.) Repeat for seven and three fish. End by placing just one fish in your pond and challenging your child to put fewer fish in her pond than are in your pond. (The only possible answer is to leave the pond empty so that there are zero fish in the pond.)

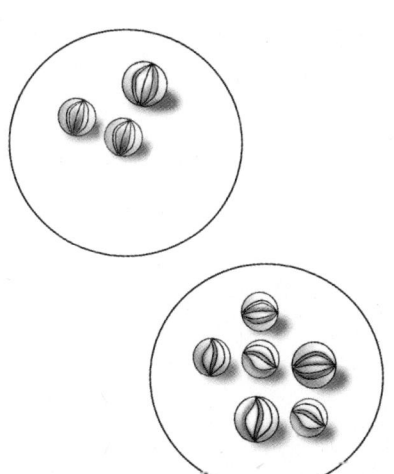

Note

Make sure to ask your child how many fish are in your pond before she adds counters to her pond so that she is explicitly thinking about the *number* of fish and not just matching her fish to yours one-by-one.

6.5 Who Has More?

Purpose

Compare quantities of concrete objects

Materials

Two small paper bags with ten counters in each bag; double ten-frames (page 101)

Activity

Have your child take a handful of counters out of one bag. Have her place the counters on the ten-frame (with one counter per box) and say how many there are. Leave the counters on the ten-frame as you take a handful of counters out of the other bag, place them on the ten-frame, and count them. Then, ask your child if she has more counters than you, fewer counters than you, or an equal number of counters.

Repeat several times. You can calibrate the difficulty level of this activity by adjusting how many counters you take. If your child is still learning to compare quantities, try to take a number of counters that is obviously different than the number your child took. If your child is ready for more challenge, try to take a number of counters that is very close to the number your child took.

Lining up the counters to see who has more. Your quantity of counters may be different than this.

Note

Young children are still learning that the size of a set does not change depending on how much space the set takes up. Take these two counter arrangements as an example. It's obvious to adults that both rows have six counters, so the two rows are equal. But many preschoolers would say that the bottom row has more counters, because that row is longer.

Using the ten-frame to compare ensures that both lines of counters are lined up with the same spacing. This helps children focus on the number of counters and not the length of the line they form.

6.6   Dice War

Purpose

Compare quantities of one to six dots without representing the quantities with concrete objects

Materials

Two regular, six-sided dice

Activity

Each player rolls a die and says how many dots are on the die. Then, have your child decide which die has more dots. If you'd like to keep score, give a point to whoever has the higher roll. Play until one player has five points or for as long as your child is interested.

Note

Comparing dots on a die is more diffi-cult than comparing concrete objects, because the dots cannot be lined up in a row and compared like counters can. However, if your child has trouble with these dot comparisons, allow her to match the dots on each die with an equal number of counters and place the coun-ters on the ten-frame to compare.

6.7 Ten-Frame War

Purpose

Use concrete objects to compare written numerals

Materials

Two sets of number cards 0–10 (pages 93-97); double ten-frames (page 101); 20 counters

Activity

Shuffle the number cards and deal the cards out face-down in two piles. When it's your turn, flip over the top card in your pile (as in the classic card game War). Place the corresponding number of counters on one of the ten-frames. Then, the other player flips over a card and places the matching number of counters on the other ten-frame. Whoever has the greater number wins both cards.

Play until you have gone through the entire deck. Then, count to see how many cards each player won. The player with the most cards wins the game.

Note

Representing the written numerals with counters helps your child understand what it means to compare two written numerals. If your child can easily compare numbers without using counters, you can skip directly to Activity 6.8 ("War Card Game").

6.8 War Card Game (Optional)

Materials

Two sets of number cards 0–10 (pages 93-97)

Activity

Shuffle the number cards and deal the cards out face-down in two piles. When it's your turn, flip over the top card in your pile. Then, the other player flips over a card. Whoever has the greater number wins both cards.

Play until you have gone through the entire deck. Then, count to see how many cards each player won. The player with the most cards wins the game.

Note

Encourage your child to think about which number comes later in the counting sequence as she decides which number is greater. If she struggles with comparing the numbers, go back to Activity 6.7 ("Ten-Frame War") and give her more practice at comparing numbers with counters on the ten-frame. It is not necessary for your child to master comparing written numerals to move on to the next chapter. She will get more experience with comparing written numerals in kindergarten.

Is My Child Ready to Move On?

Your child is ready to move on when she can compare any two numbers between zero and ten, either by representing the two numbers with counters or by comparing the written numerals directly.

— CHAPTER 7 —
ADDITION AND SUBTRACTION STORIES

Chapter Overview

Chapter 7 focuses on understanding the concepts of addition and subtraction. You'll use simple stories, concrete objects, and pictures to introduce addition as joining two sets together, and subtraction as taking away part of a set.

7.1 Penny Addition Stories, Part One

Purpose

Introduce addition as joining two groups

Materials

Five pennies

Activity

Tell your child the following addition story: "I had three pennies." (Take three pennies and put them in front of you in a line.) "Then I got two more pennies." (Add two more pennies to the line, leaving a space between the two groups.) "How many pennies do I have now?" (Five.)

Repeat this activity with other penny addition stories. For example, you might tell a story about starting with three pennies and earning one more (for a total of four) or starting with two pennies and then finding one more (for a total of three). Stick with small quantities for now so that the focus is on the concept of joining the two groups and not the calculations. Also include zero in some of the stories. (For example, you might say, "I had three pennies but then I didn't get any more. How many did I have then?")

Note

Your child learned to recognize amounts up to five by sight in Chapter 3. However, it's not unusual for children to revert to counting one-by-one when presented with an unfamiliar problem. Gently encourage your child to recognize the sums rather than counting them one-by-one in this activity, but be patient if she needs to count one-by-one at first.

7.2 Penny Addition Stories, Part Two

Purpose

Practice addition with concrete objects

Materials

Seven pennies

Activity

Tell this penny addition story: "I had four pennies, and then I got three more." Ask your child to act out the story with pennies. He should make a group of four pennies, then make another group of three pennies. Ask him how many pennies there are altogether. (Seven.)

Then, ask your child to tell a penny story and act it out with the pennies. For example, he might say, "I had three pennies and then I found one more. Now I have four pennies."

Ask your child to tell several different penny stories. If he's stuck, give him a number to start with: for example, you

might say, "How about if you start your story with two pennies and then you earn some more?" Continue to use only small numbers so that the focus is on the idea of joining the two groups and not the calculations.

7.3   Add Handfuls

Purpose

Practice addition with concrete objects

Materials

Ten counters

Activity

Put three counters in your right hand and four counters in your left hand. Show both amounts to your child and ask, "How many counters do I have altogether?" (Seven.) Encourage your child to count on from four to find the total ("Four...five...six...seven!") but allow him to count all of the counters one-by-one if necessary.

Repeat the activity with different numbers of counters whose sum is less than ten. Make sure to occasionally include zero (with an empty hand). Also reverse the activity: have your child take two handfuls of counters and ask you the total.

Note

Like the penny stories in Activities 7.1 and 7.2, this activity helps your child understand addition as joining two groups. However, the penny stories and handfuls of counters present two slightly different joining situations. In the penny stories,

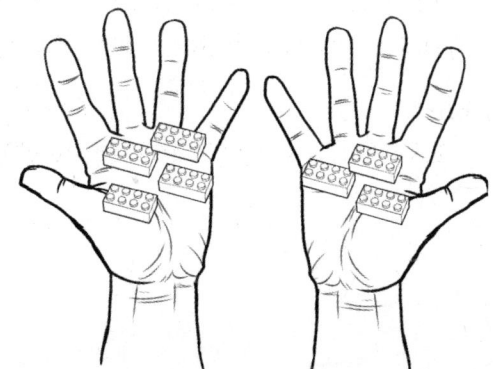

the joining is dynamic: you begin with some pennies and then add on a new group. With the handfuls of counters, the joining is static: your child sees both groups of counters at the beginning and mentally joins the two groups together to find the total. Real life presents both types of joining situations, so children need to understand both.

7.4   Fish Pond Addition Stories

Purpose
Use pictures to represent addition

Materials
Blank paper; two different-colored writing utensils

Activity

On a piece of paper, draw a large circle to be the "fish pond." In the pond, draw two fish of one color and three fish of another color.

Ask how many fish are in the pond. (Five.)

Repeat with other small numbers of different-colored fish. For example, if you have a blue marker and a green marker, you might draw one blue fish and four green fish (for a total of five), or you might draw two blue fish and no green fish (for a total of two).

Note
Like Activity 7.3 ("Add Handfuls"), this activity presents a static joining situation.

7.5   Penny Subtraction Stories

Purpose

Introduce subtraction as taking away part of a set

Materials

Five pennies

Activity

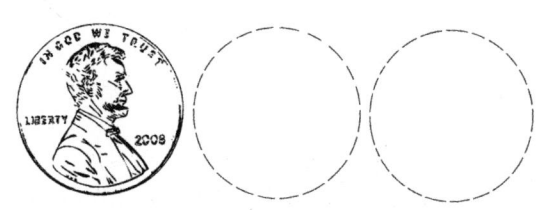

Tell your child the following subtraction story: "I had
three pennies." (Take three
pennies and put them in
front of you in a line.) "Then
I lost two of my pennies." (Remove two pennies.) "How many pennies do
I have now?" (One.)

Repeat this activity with other subtraction stories. For example, you
might tell stories about starting with four pennies and giving two away
or starting with five pennies and spending three of them. As you did with
addition, use only small quantities for now so that the focus is on understanding the concept of subtraction and not calculations.

7.6   Subtraction Snack

Purpose

Practice subtraction with concrete objects

Materials

Single ten-frame (page 99); ten small snacks (crackers, cereal, raisins,
etc.) (If you prefer not to use food items, you can use counters instead.)

Activity

Have your child place one snack in each box of the ten-frame. Ask how many snacks there are. (Ten.)

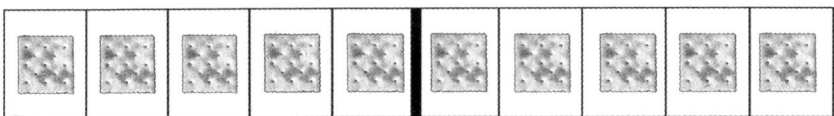

Tell your child that you are going to pretend that he is a big hungry giant and that his job is to help you act out a subtraction story. (If your child doesn't like giants, you can instead suggest being a dinosaur, princess, or monkey—whatever your child enjoys pretending to be.)

As you tell the following story, have your child pretend to be the giant and eat the number of snacks mentioned in each part of the story. Make sure he starts at the right side of the ten-frame and removes each snack in a row.

"Once upon a time there was a big hungry giant. He had ten snacks, but then he gobbled up three of the snacks. How many did he have left?" (Seven.)

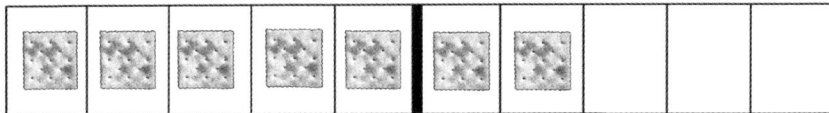

"Then, the giant was still hungry, so he ate two more snacks. How many were left then?" (Five.)

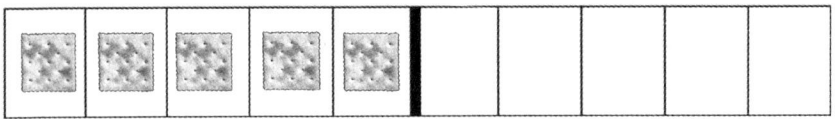

"The giant was getting pretty full, so he decided to eat zero snacks. How many were left then?" (Still five.)

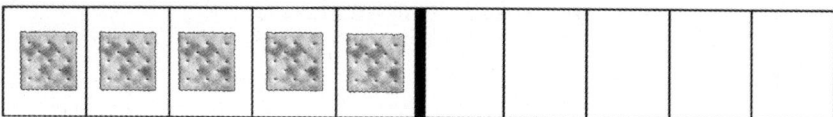

"After the giant took a nap, he was hungry again. This time, he ate four snacks. How many did he have left now?" (One.)

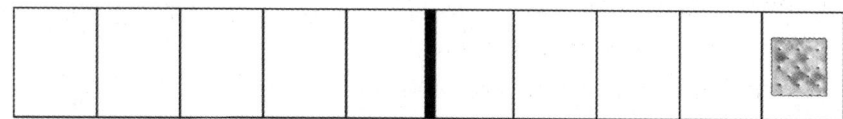

"Finally, the giant polished off the food and ate the last snack. How many did he have left then?" (Zero.)

Note

If your child enjoys this activity, repeat it at snack time. You can vary it by using different types of snacks and different starting numbers less than ten.

7.7   Balloon Subtraction Stories

Purpose
Use pictures to represent subtraction equations

Materials
Blank paper; writing utensil

Activity

On a piece of paper, draw five simple balloons. Tell your child this subtraction story: "I had five balloons, but then two of them popped." Cross out two of the balloons to show the ones that popped. Ask how many balloons were left. (Three.)

Repeat with other small numbers of balloons. For example, you might draw four balloons and say that one popped, or you might draw five balloons and say that all five popped.

Then have your child tell some balloon stories and draw pictures to match.

Is My Child Ready to Move On?

Your child is ready to move on when he understands the concept of addition as joining two sets and subtraction as taking away part of a set. His next step should be a kindergarten math program that teaches him how to count to 100, understand place-value, identify simple shapes, and use coins, along with providing much more practice with addition and subtraction.

APPENDIX

Five-Frame

You will use the five-frame in Chapter 3 to help your child learn to recognize quantities up to five.

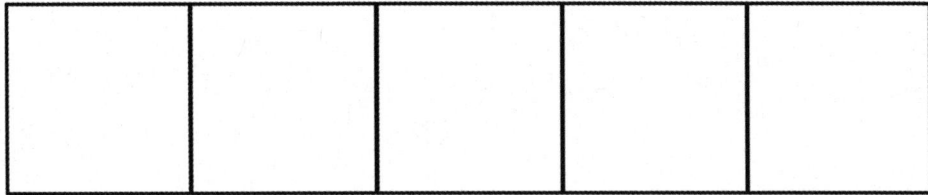

Number Race Game Board

You will use this game board to play Number Race in Chapter 5.

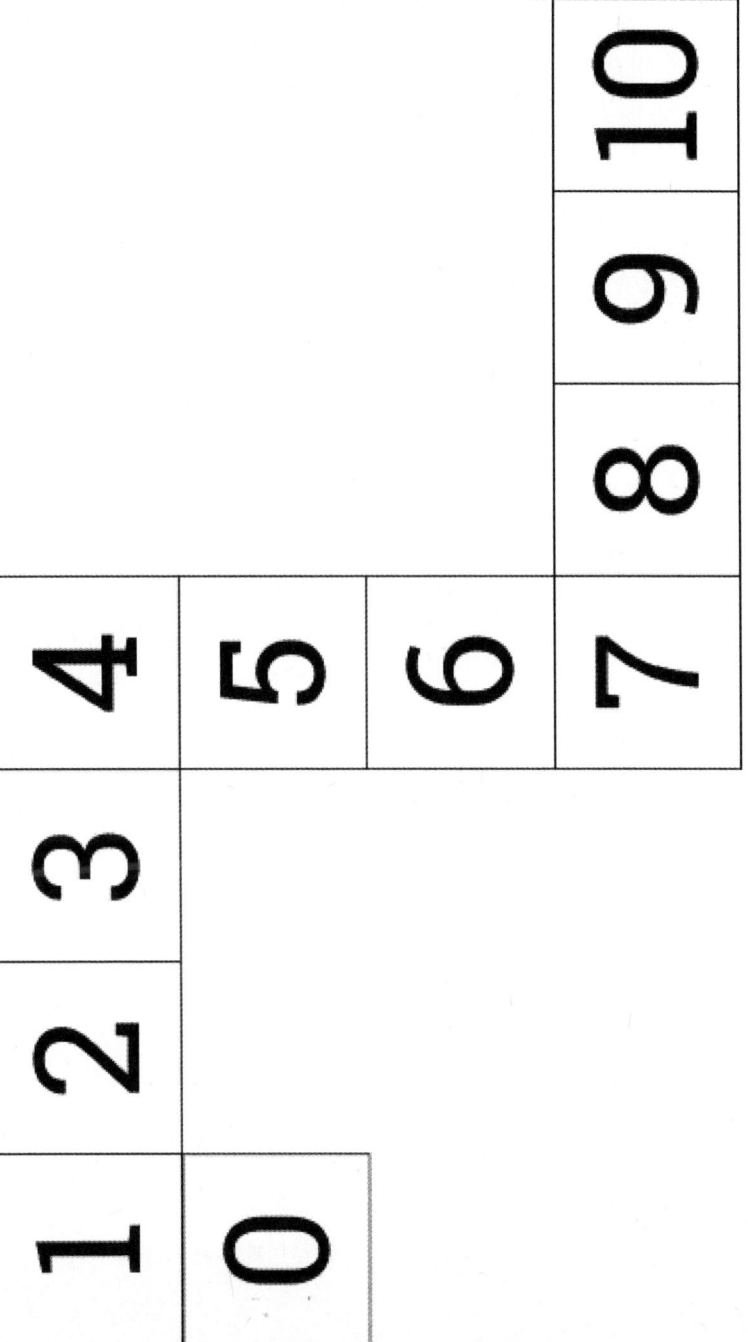

Number Cards

You will use the number cards to introduce your child to written numerals in chapter 5. You will also use the number cards to compare numbers in Chapter 6.

Directions: Cut out the number cards on this page and the following two pages so that you have two sets of the numbers 0–10. (For sturdier cards, copy them onto cardstock first.)

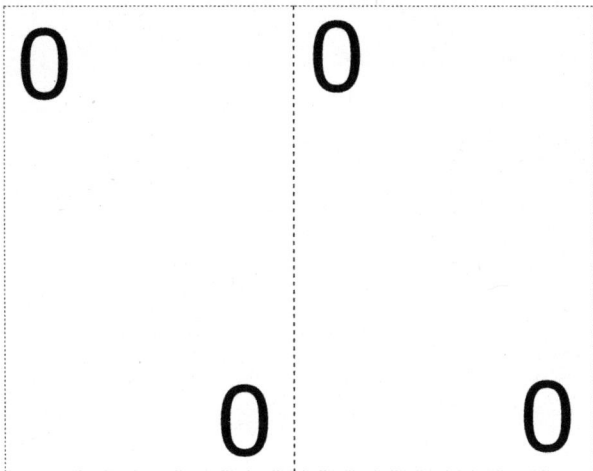

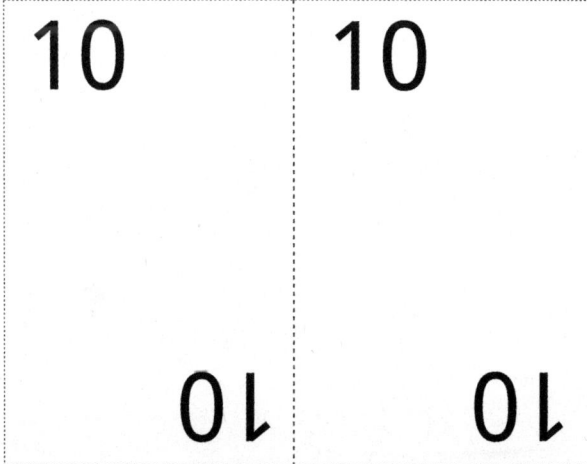

Number Cards

1	2	3
1	2	3

4	5	6
4	5	6

7	8	9
7	8	9

Number Cards

1	2	3
⇂	乙	Ɛ
4	5	6
ㄣ	5	9
7	8	9
ㄥ	8	6

Single Ten-Frame

You will use the single ten-frame in chapters 4 and 7 as your child explores the numbers from six to ten.

Double Ten-Frame

You will use the double ten-frames in chapters 4 and 7 as your child learns to understand and compare numbers up to ten.